MATHS MAGIC

# Charts and Graphs

Written by Wendy Clemson
and David Clemson

www.two-canpublishing.com

Published by Two-Can Publishing
43-45 Dorset Street, London W1U 7NA

www.two-canpublishing.com



For information on Two-Can books and multimedia,
call (0)20 7224 2440, fax (0)20 7224 7005
or visit our website at http://www.two-canpublishing.com

Created by
act-two
346 Old Street
London EC1V 9RB

www.act-two.com

**Authors:** Wendy Clemson and David Clemson

**Editor:** Penny Smith
**Designers:** Maggi Howells, Helen Holmes and Liz Adcock
**Illustrators:** Andy Peters and Mike Stones
**Photographer:** Daniel Pangbourne
**Pre-press production:** Adam Wilde



'Two-Can' is a trademark of Two-Can Publishing.
Two-Can Publishing is a division of Zenith Entertainment plc,
43-45 Dorset Street, London W1U 7NA

Hardback ISBN 1-85434-880-9
Paperback ISBN 1-85434-881-7

Dewey Decimal Classification 511

Hardback 10 9 8 7 6 5 4 3 2 1
Paperback 10 9 8 7 6 5 4 3 2 1

A catalogue record for this book is available
from the British Library.

Colour reproduction by Colourscan Overseas, Singapore
Printed in Hong Kong

# Contents

# Chart or graph?

There is so much information around us that sometimes it's hard to take it all in! To help, you can organise it into lists, charts, tables and graphs. This makes information easier to understand. All the information in this picture can be turned into charts and graphs that you can find out about in this book.

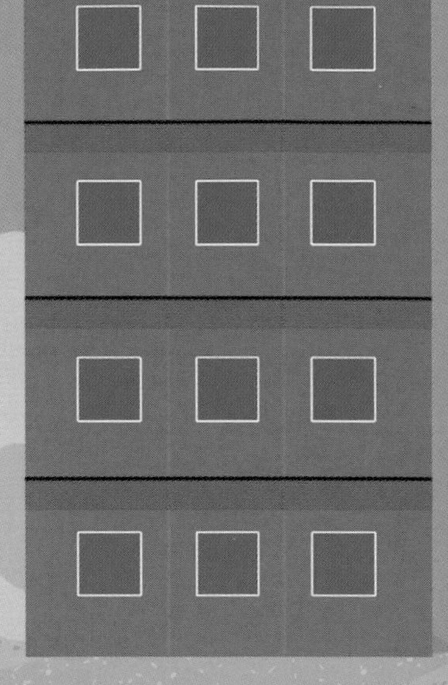

| | |
|---|---|
| Bristol | 192 km |
| Cardiff | 243 km |
| Glasgow | 654 km |
| Southampton | 128 km |

1. The places on this sign are in the order of the alphabet. If you place them in order of distance, starting with Southampton, which comes second in the list?

2. Look at the children playing football. One group is wearing red clothes. Another group is wearing blue. What is the third group wearing?

3. There are three families living on each floor of these blocks of flats. Which block has the most families?
4. Lots of cars are on the road today! How many can you see with black tyres and a roof rack? Now count the cars with black tyres and no roof rack.
Answers: 1) Bristol 2) red and blue 3) the yellow block 4) one, three

# On the list!

One of the simplest ways to organise information is to make a list. Our chef has a busy day ahead. He has to decide what to put on his menu, buy all the right ingredients, then make the meals. Lists help him to manage. Let's see if you can help him too!

## FRUIT SALAD RECIPE

6 oranges
6 bananas
4 pears
2 kiwi fruit
1 pineapple
1 bunch of grapes
1 punnet of strawberries

Squeeze two of the oranges into a bowl. Chop up the rest of the fruit, mix it with the orange juice, then serve.

### Chef's busy day

**1** The chef has found a great recipe for fruit salad. He puts it on the menu, then writes a shopping list. Check the shopping list against the recipe. Which fruit has he forgotten to include?

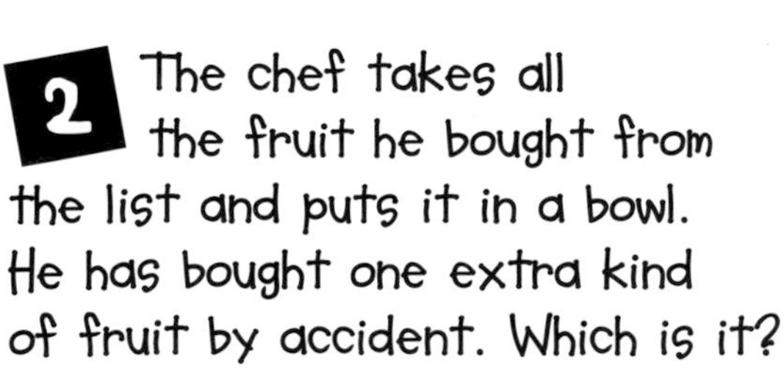

**2** The chef takes all the fruit he bought from the list and puts it in a bowl. He has bought one extra kind of fruit by accident. Which is it?

## TODAY'S MENU

Cheese sandwich £1.50
Chicken pie £2.00

Fruit salad £2.50
Ice cream £2.00

Milk £1.00
Fruit juice £1.20

**3** One of the chef's customers orders the meal below. Find the prices listed on today's menu, then add them together to work out how much the meal costs.

## PROVE IT!

The order of a list can be important. Look at the list of instructions below. It tells you how to make a sandwich. But the list is in the wrong order, so it's impossible to make the snack.

- Cover filling with second slice of bread.
- First cut loaf into slices.
- Arrange filling on one slice of bread.

Put the list in the right order by matching it to the pictures. Now it's easy to make the sandwich!

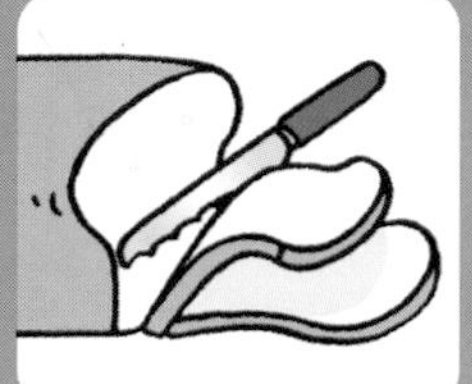

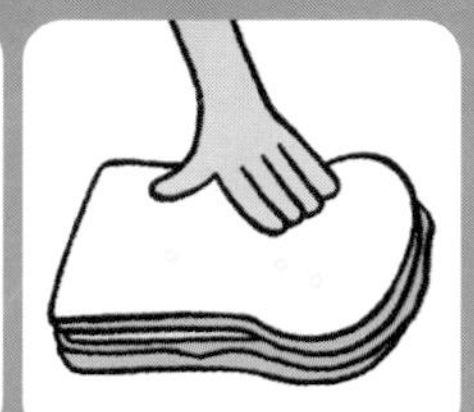

Now try this

## BRAIN teaser

Challenge a friend and put her memory skills to the test.

First lay 12 things untidily on a tray. Ask your friend to look at them for a count of five. Then put the tray out of sight. How many things can she remember?

Play the game again using different items, but lay them out in order, like a list, with similar things next to each other. Your friend will remember more things this time.

Answers: 1) two kiwi fruit 2) apples 3) £5.20

# Table time

When you have lots of information, making a table is often a good idea! A table is particularly useful for comparing two or more sets of information. Play the ocean board game and make a table to keep score. Compare your final scores to find the winner.

### Make an ocean board game

**YOU WILL NEED**
coloured pens, stiff coloured card, scissors, craft knife, 10 corks, paints, sticky tape, glue, table-tennis ball

**1** First trace the shark's head on the left on to card. Cut it out and draw around it to make ten sharks. Decorate them and cut them out.

**2** Paint 10 corks, then ask an adult to cut a slit in the top of each one. Push a shark's head into each slit.

**3** Take a strip of card, 80 cm long and 7 cm wide. Fold it into the concertina shape shown below.

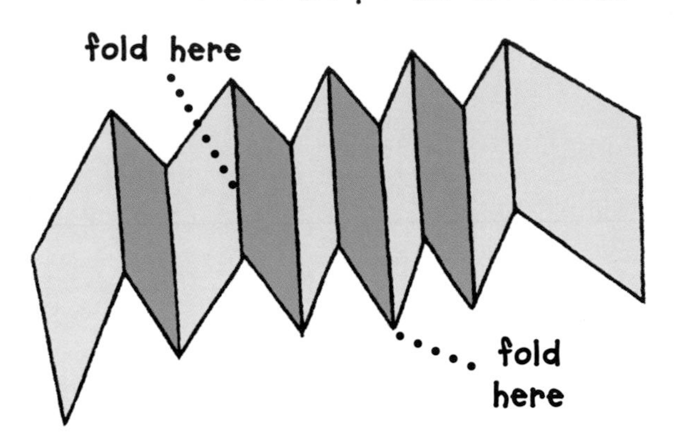


**4** Cut out an oblong of card, 60 cm long and 40 cm wide. Tape the concertina to the card, write in the numbers and mark the crosses in the positions shown. Glue a cork shark to each cross.

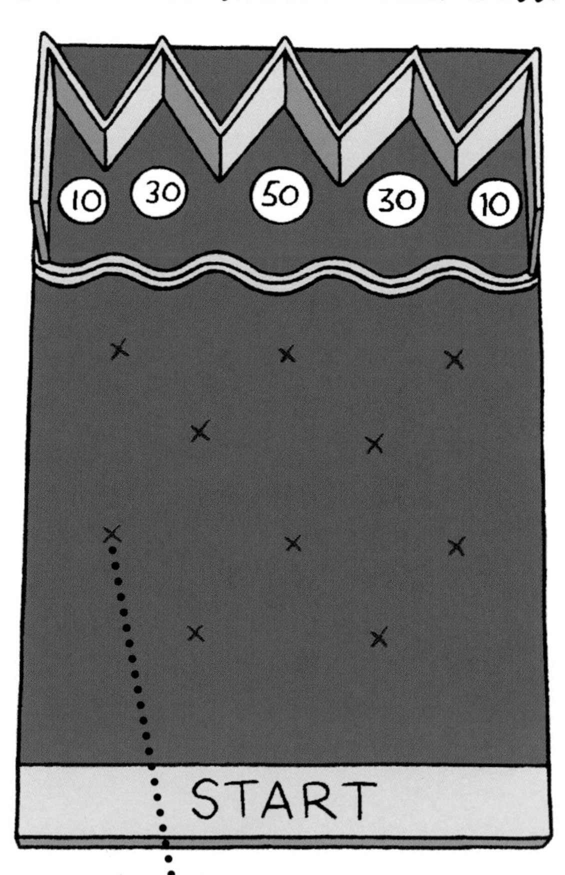


Red crosses show where to glue your cork sharks.

**5** Now draw up a score table like the one below, but don't fill in the scores. Read the instructions on the right, then play the game!

Write the number of rolls in this column.

Write the players' names at the top of the columns.

SCORE TABLE

| | Tom | Susy |
|---|---|---|
| First roll | 30 | 30 |
| Second roll | 10 | 50 |
| Third roll | | |
| Fourth roll | | |
| Total score | | |

Write your score under your name after each roll.

Add up the numbers in each column to find each player's total score.

### HOW TO PLAY AND KEEP SCORE

- ◆ This game is for two or more players. Keep track of your scores on the score table.
- ◆ Take turns rolling the table-tennis ball down the slope into the different sections.
- ◆ Each time you roll, write down the number of points you scored under the column with your name.
- ◆ Have four rolls each, then add up your totals. The person with the highest score is the winner!

It's my turn to roll the ball.

START

Prop up your board with magazines or books.

# Calendars

This calendar page is for the month of August. There are 31 days in this month.

How do you remember important dates, such as birthdays? It's easy with a chart called a calendar. On a calendar, there is a page for each month. You can also find every day and date of the year. Look at the calendar page on the right, check out how it works, then help Rosie to answer the questions.

Dates run from side to side in rows.

## AUGUST

| Monday | Tuesday | Wednesday | Thursda |
|---|---|---|---|
| | | | 1 |
| 5 | 6 | 7 | 8 |
| 12 | 13 | 14 | 15 |
| 19 | 20 | 21 | 22 |
| 26 | 27 | 28 | 29 |

### Rosie's important dates

**1** Rosie wants to learn to swim and is about to start lessons. They begin on the first Thursday of the month. What date is that?

**2** Her birthday is on the 27th August. What day is this?

**3** Rosie's mum has to work on the second weekend of the month. What are the dates?

The days of the week are shown at the top.

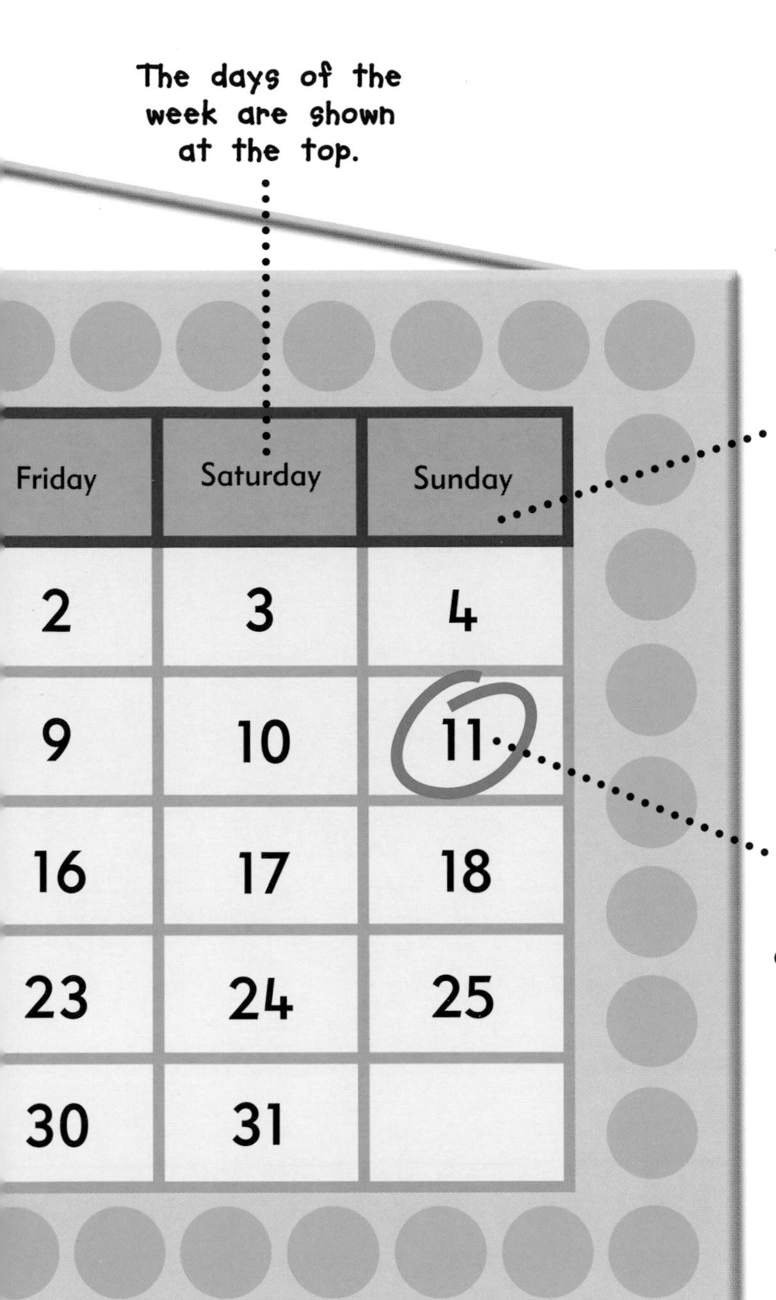


Columns run up and down the calendar. To find out on which day of the week a date falls, look at the top of a column.

The 11th of August falls on a Sunday.

## HOW MANY DAYS?

Here's a clever rhyme to help you remember how many days there are in each month.

30 days have September,
April, June and November.
All the rest have 31,
Except February alone,
And that has 28 days clear,
And 29 in each leap year.

## PROVE IT!

You can tell when it's a leap year by looking at the last two digits, or numbers, of that year.

2004 .... last two digits

If you can divide the digits exactly by 4, then it's a leap year. Years that end in 00, such as 2000, are only leap years if they can be divided by 400.

Answers: 1) 1st August 2) Tuesday 3) 10th and 11th August Prove it!) 2016

# Pictograms

A pictogram is a chart that shows information in pictures, called symbols, rather than numbers. Each symbol stands for a particular number of objects. Now try making a pictogram of the animals in the picture of the safari park.

### Make a safari pictogram

**YOU WILL NEED**
felt pen, washing-up sponge, scissors, paint, plate, paper

**1** Trace this paw print symbol on to a piece of sponge, then cut it out.

**2** Pour paint on to a plate. Dip one side of your paw print into the paint. Then press it on to paper. Practise this a few times.

**3** Now make a chart like the one below. Fill it in, using one paw print to stand for each animal in the safari park. The first row is done for you.

## Count the animals in twos

The safari park has a special petting area. We've made a pictogram to show the animals in the petting area. This time each paw print stands for **two** animals.

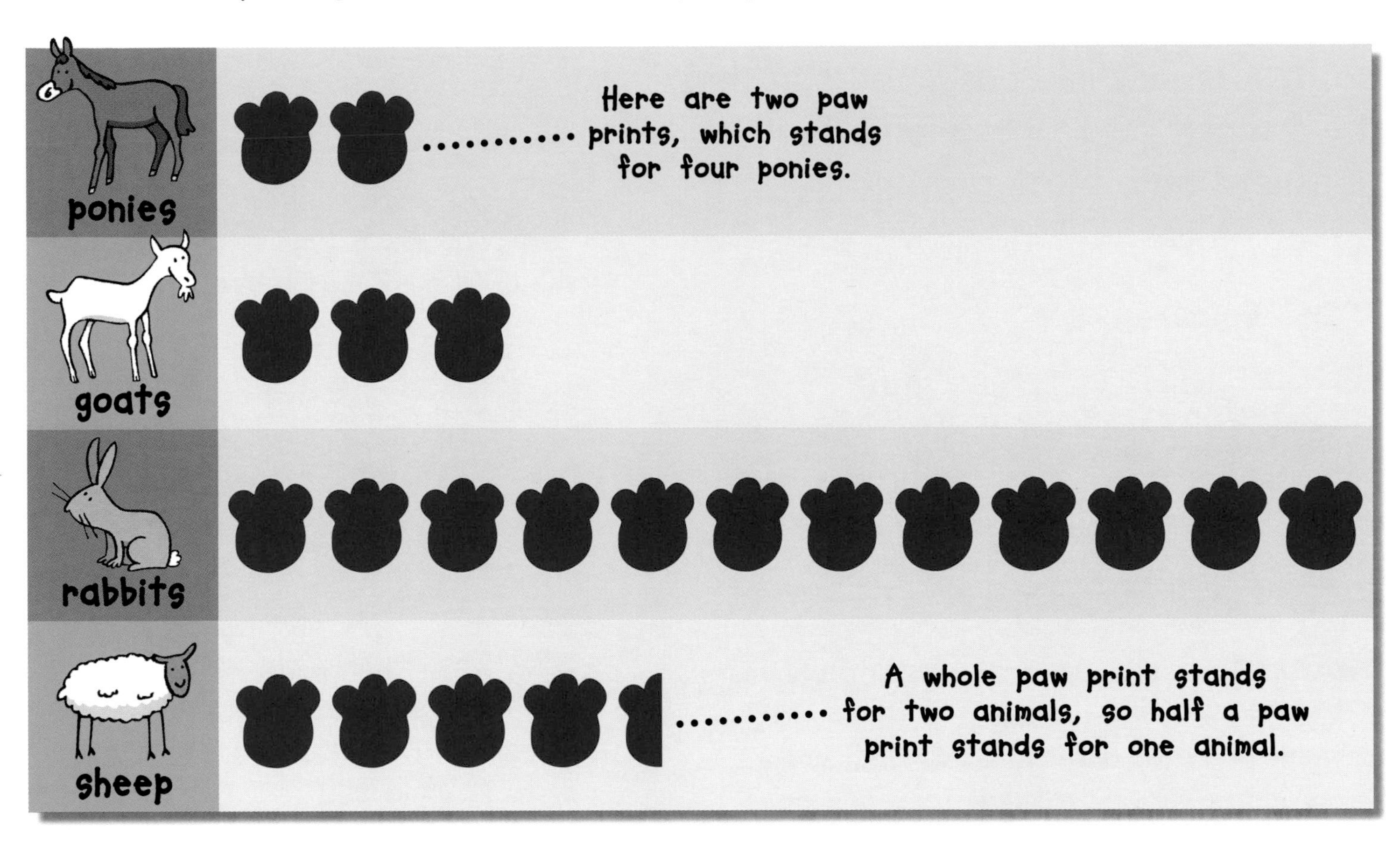


**1** How many goats live in the petting area?

**2** How many more sheep than goats are there?

**3** There are 24 of one type of animal. Can you tell which animal this is?

**4** If the goats have four babies between them, how many paw prints need to be added?

### Now try this BRAIN teaser

Tom and Gus are giant tortoises. Each day, Tom eats six lettuces and Gus eats nine. Each symbol stands for three lettuces.

Challenge a friend to say where the five lettuce symbols go on this pictogram.

Gulp! Here's the finished pictogram. Did your friend get the answer right?

Answers: 1) 6 goats 2) 3 more sheep 3) rabbit 4) 2 more paw prints

# Block graphs

What's your favourite pet? Which pets do your friends like best? You can show the answers to all these important questions on a block graph!

## Cat colours

**1** There are nine crafty cats above. Four cats are ginger, three are black and two are white. On the graph, one square block stands for each cat.

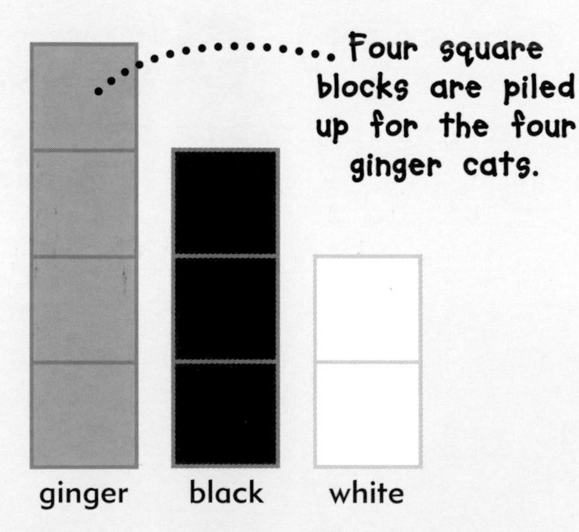


**2** There are separate piles for each cat colour. Now count the number of ginger cats, black cats and white cats.

## Make a block graph of favourite pets

**YOU WILL NEED**
pen, paper, white card, ruler, scissors, glue, paints

**1** Ask each of your friends to tell you which type of animal he or she likes best. Write down the answers.

**2** Draw a grid on card. You need one square for every answer given. So, if you asked 15 friends, draw 15 squares. Make the squares the same size and cut them out.

**3** Now colour your squares to show your answers. If five friends said they liked cats best, colour five squares green, and so on.

**4** Stick the squares on a larger piece of card to make your block graph. Label the graph and give it a title. On our graph below, which is the most popular pet?

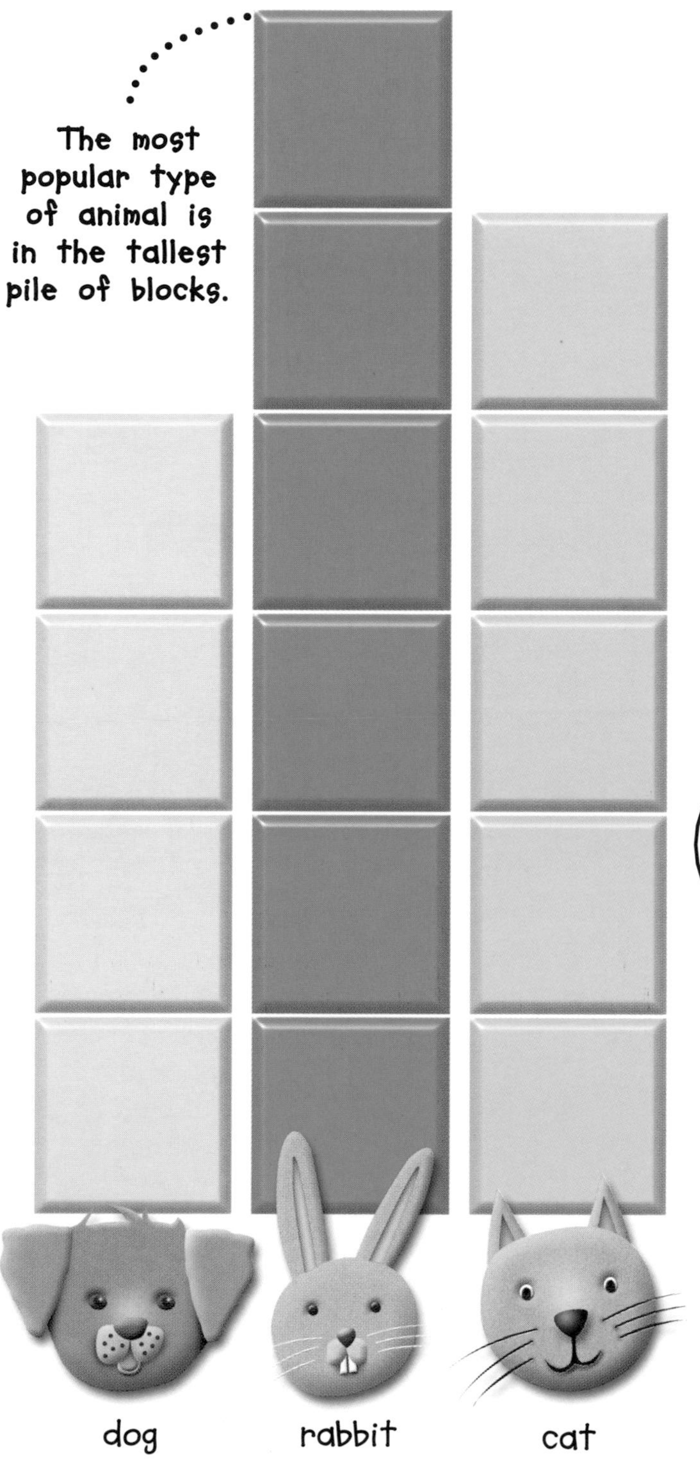


Now try this

## BRAIN teaser

Seven cats were shown three different foods. Challenge a friend to work out from this block graph how many cats did not choose fish.

To find the answer, count the blocks for beef and lamb instead of fish. There are six blocks, so six cats did not choose fish.

beef lamb fish

# Bar Charts

A bar chart is made up of bars of different heights. Along the bottom, there is a line called an axis. Another axis runs up the side of the chart. Check out how a bar chart works below. How many children are wearing shorts and how many are wearing trousers?

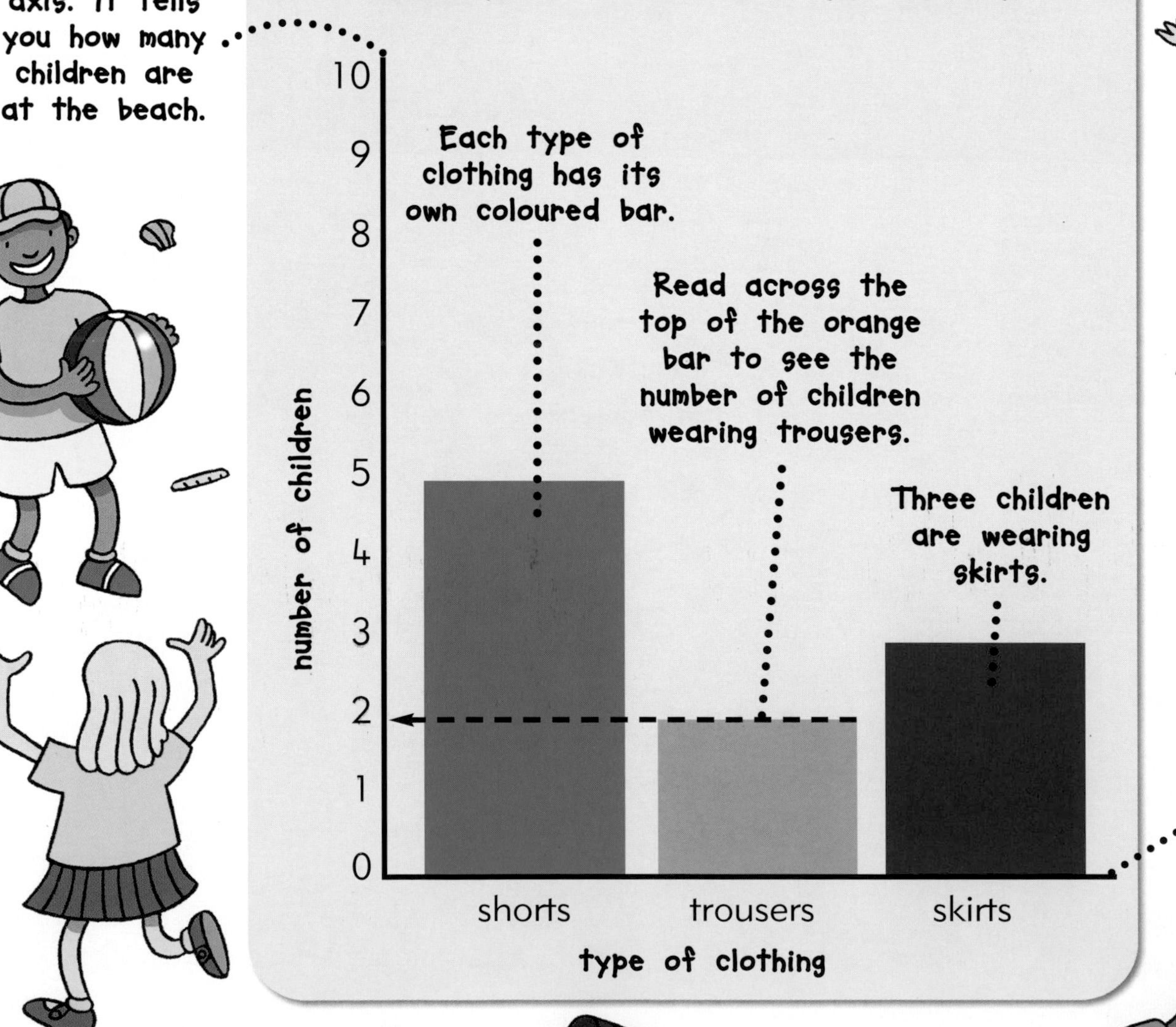


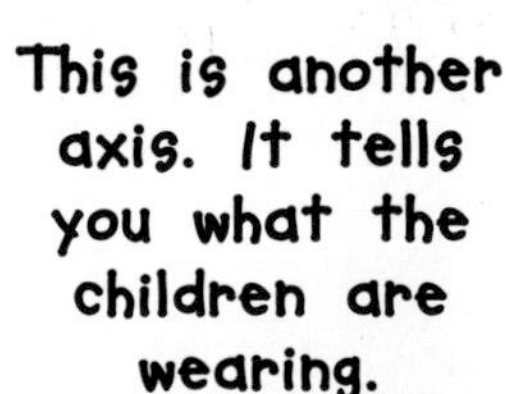

## Top tip

Make reading a bar chart easy by holding a ruler along the top of a bar to the numbers on the axis. Read straight across.

### Safe surfing

These charts show the number of days you can surf from June to September at two different beaches.

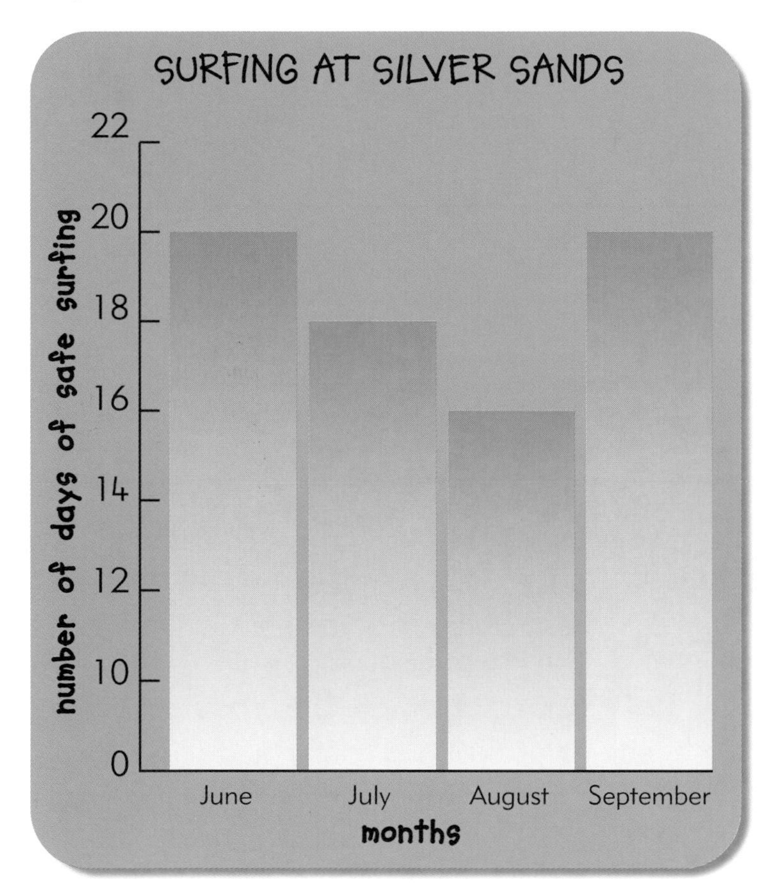


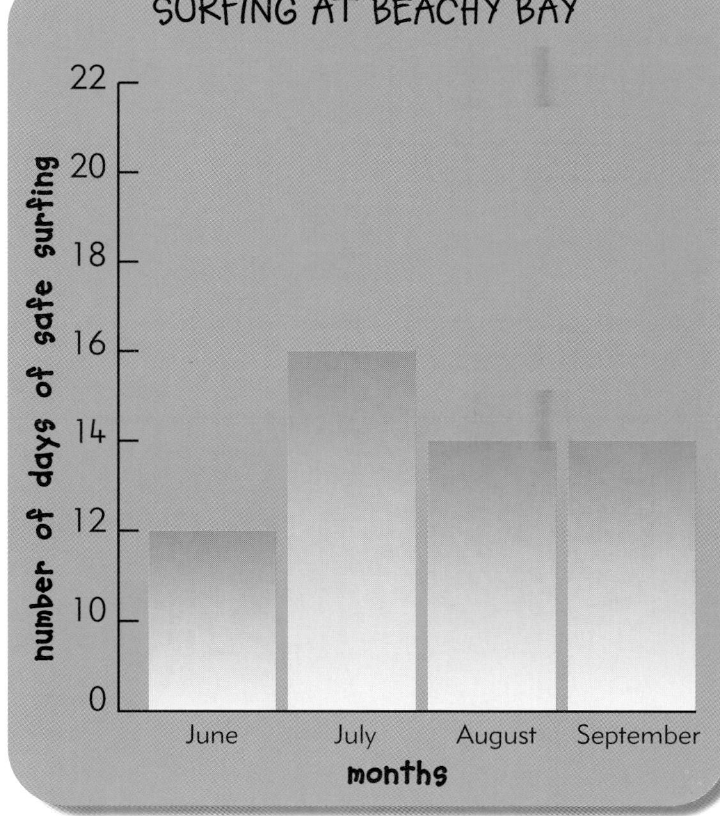


**1** Look at the bar chart for Silver Sands. How many days would you expect to surf there in August?

**2** Now look at the bar chart for Beachy Bay. In which month can you surf the least?

**3** In the month of July, which beach would you visit for more days of surfing?

**4** In which months can you surf for 20 days?

Answers: 1) 16 days 2) June 3) Silver Sands 4) June and September at Silver Sands

# Venn diagrams

Look at the big picture. Similar objects are grouped together. All the red shapes are in one group, or set, and all the triangles are in another. A few triangles are red, so they fit into both sets and are placed where the sets overlap. The picture is called a Venn diagram.

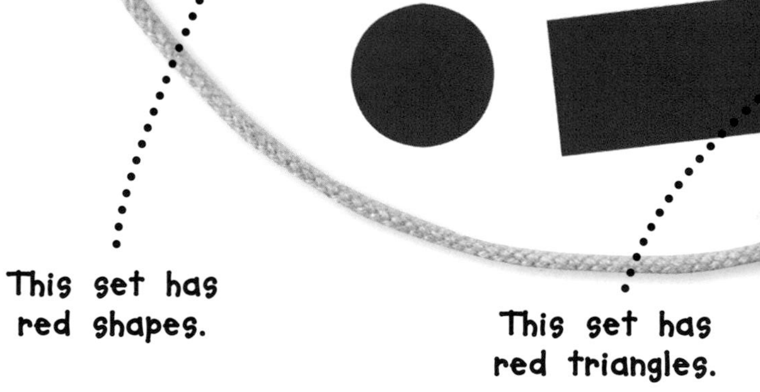


**Make your own Venn diagram**

**YOU WILL NEED**
card, scissors, string, different coloured pens

**1** First make 16 small card shapes. Cut out eight squares, three triangles, three oblongs and two circles.

**2** Colour your triangles, oblongs and circles red. Now colour your squares. Make four of these red and the rest different colours, such as yellow, blue and green.

**3** Cut two long pieces of string. Knot together the ends of each piece to make two loops. Lay the loops on the table, making sure they overlap.

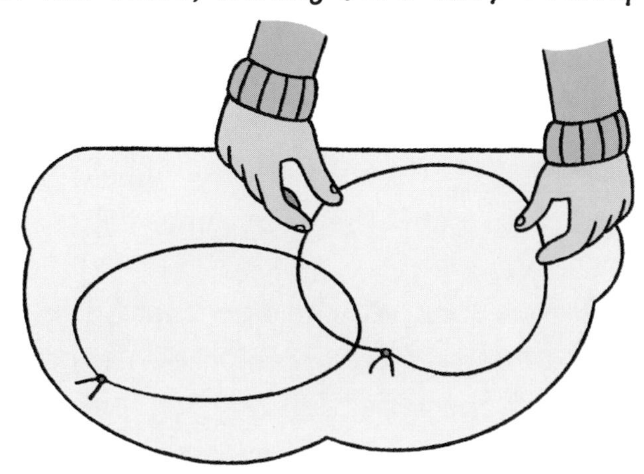

**4** Sort your shapes into sets of squares and red shapes. How many red shapes are there? Which shapes do you have in your overlapping set?

## Now try this

# BRAIN teaser

Challenge a friend to tell you how the animals have been sorted in this Venn diagram.

Here's a clue. Think about where these animals live.

The answer is one set of animals lives in water. One set lives on land. The middle set lives both in water and on land. Did your friend work out the answer?

Answers to question 4: 12 red shapes, red squares

# Carroll diagrams

On this page, we sort aliens that are happy or sad and aliens that are red or blue. When you sort objects into groups, they may have more than one similarity. You can use a Carroll diagram to compare different kinds of information at the same time. Now make your own alien Carroll diagram.

## Make an alien Carroll diagram

**YOU WILL NEED**
coloured paper, pencil, felt pens, scissors, coloured card

**1** First fold a strip of paper, 7 cm wide by 42 cm long, backwards and forwards 9 times to make a concertina. Close up the concertina and copy one alien outline from below on the front.

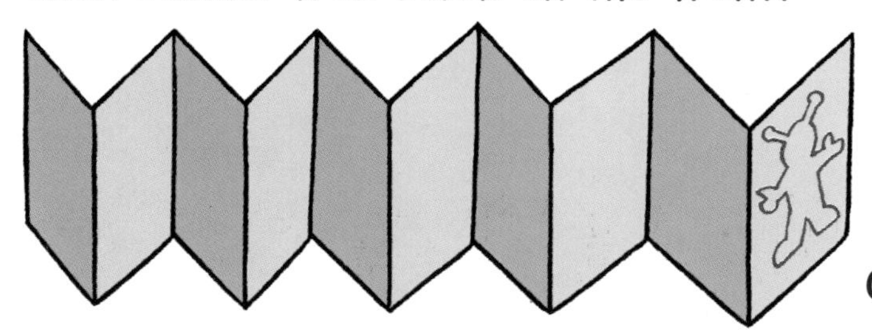

**2** Now cut through all the layers of paper to make 10 aliens. Colour your aliens and draw faces, making sure they exactly match the ones at the bottom of the page.

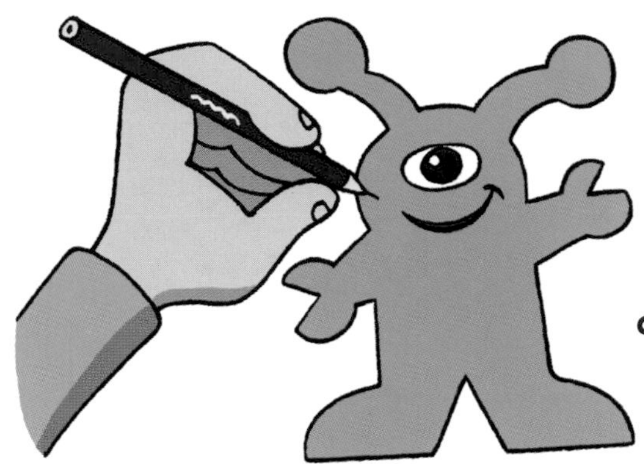

This is a blue alien with one eye and a big smile.

**3** Next, make your board by copying the one below on to a piece of card. Make each of the four big squares 15 cm long and 15 cm wide.

Copy the outline carefully.

**4** Write your headings on the board in pencil as shown, then sort out your aliens.

A column runs up and down. Place your happy aliens in this column.

**5** Now change the column headings to 'one eye' and 'two eyes'. Sort your aliens again. How many blue aliens with one eye can you count?

**6** Finally, cut off the antennae on five of your aliens and try to make a new diagram showing happy and sad aliens with and without antennae.

Answer to question 5: three blue aliens with one eye

# Tally Charts

Keep score in this high-jumping cross-country horse race by using a tally chart. In a tally chart, you count in fives. Marks stand for numbers like this:

| for one    || for two,

||| for three    |||| for four

To write the number five, draw a line through the four strokes ~~||||~~ . Then start counting again. See how easy it is to add up your scores!

~~||||~~   ~~||||~~   |||

5 + 5 + 3 = 13

**RULES OF THE GAME**

- This is a game for 2 to 4 players.
- Take turns throwing the dice, then move the counter the number of places shown on the dice. You can move forwards or backwards but not diagonally.
- Jump over an obstacle to win points as follows:

- When you reach 50 points, gallop to the finish. The first person to reach there is the winner. Tally ho!

## Play the Tally-Ho! game

**YOU WILL NEED**
coloured card, felt pens, scissors, glue, dice

**1** To record your scores, make a tally chart like the one below. Ours is already filled in.

| | |
|---|---|
| Joseph | ~~||||~~ ~~||||~~ ~~||||~~ || (17) |
| Kate | ~~||||~~ ~~||||~~ ~~||||~~ ~~||||~~ (20) |

**2** Then, make one counter for each player. Copy this template on to card and cut it out. Cut along the line to the centre of the circle and glue the edges together.

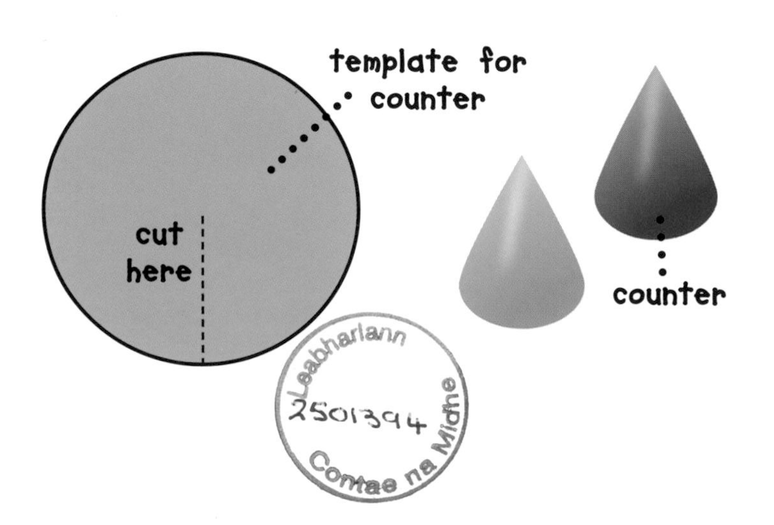

START
FINISH

# Frequency tables

It's time for touchdown! First make a frequency table, then put your skills to the test with our airport activity. A frequency table shows you how many times a particular event happens. In the activity, does one colour plane land more frequently than another?

## Make a frequency table

**YOU WILL NEED**
four different coloured beads, small bag, pen, paper

**1** Put the four beads into a bag. Make sure the beads are all the same size and shape.

**2** Without looking, take out a bead. Write down its colour, then put the bead back and shake the bag. Do this 20 times. Keep a tally and record your results in a table, like this one.

| Colour | Frequency | Total |
|---|---|---|
| Red | 𝍸 I | 6 |
| Blue | III | 3 |
| Green | IIII | 4 |
| Yellow | 𝍸 II | 7 |

We picked the red bead six times.

You may have different numbers.

## At the airport

**YOU WILL NEED**
card, coloured pens, one red bead, one blue bead, one green bead and one yellow bead, small bag, paper

**1** First make your planes. Trace the one here on to card. Then cut it out.

**2** Draw around your plane to make 20 planes altogether. Colour five red, five blue, five yellow and five green. Then cut them out.

Decorate your planes with felt-tip pens.

**3** Now make your runways. Cut out a piece of card 28 cm by 50 cm. Divide it into four long strips as shown and draw lights along each one. Lay all the runways on a table.

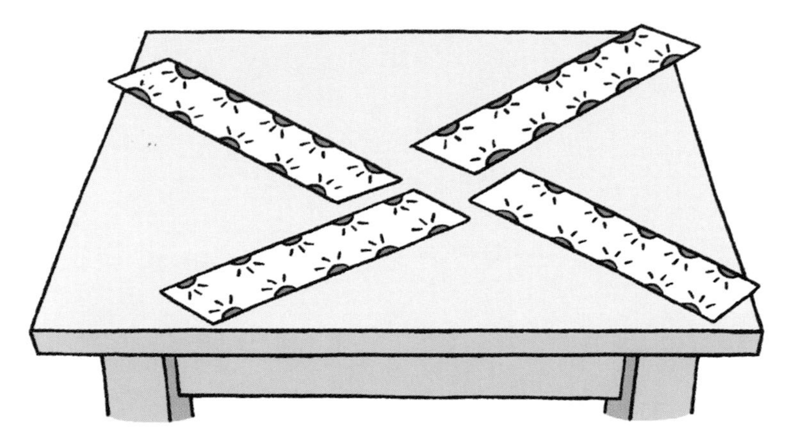

**4** The aim of the activity is to land five planes of one colour on a runway. To decide which colour plane you can land, pick beads out of a bag like you did on the opposite page.

**5** Now draw up a frequency table like the one below and fill in your results. Your results may be different from ours.

| Colour of plane | Number of planes on runway | | | |
|---|---|---|---|---|
| | Round 1 | Round 2 | Round 3 | Total number of planes |
| Red | 2 | | | |
| Blue | 1 | | | |
| Green | 1 | | | |
| Yellow | 5 | | | |

**6** Carry out the activity two more times, filling in your table each time. Then add up the total number of red, blue, green and yellow planes. Which colour plane is on the runway most frequently?

**WOW!**

All four beads have an equal chance of being picked. But you may need to play the game hundreds of times before the number of colours are equal on the frequency tables!

# Choose a Chart

You can often turn information into more than one kind of chart. But the trick is to choose the chart that shows the information most clearly and is quickest to read. Take a closer look at the charts on these pages. They tell you about the tropical fish in the picture. Which charts do you think work best?

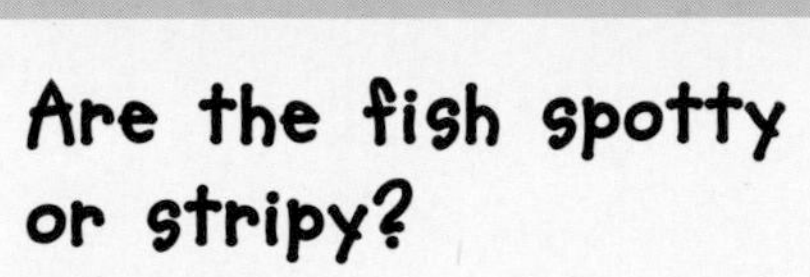

## Are the fish spotty or stripy?

These charts show which fish have only stripes, which have only spots and which have both spots and stripes. Look at the picture to answer the questions, then check that each chart is correct.

**1** How many fish have only spots?

**List**

Three fish have only stripes

Four fish have only spots

Two fish have both spots and stripes

**2** How many fish have both spots and stripes?

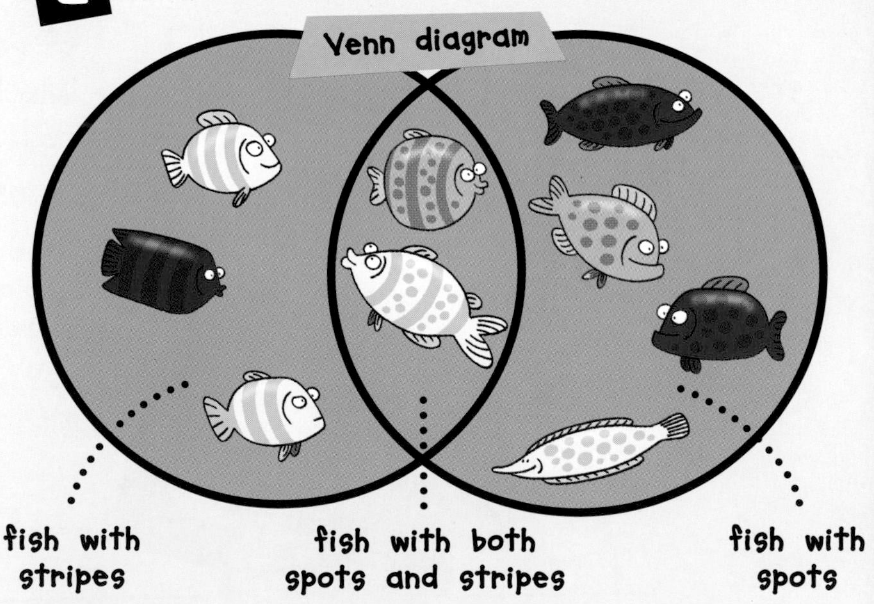


**3** Now look at both the list and the Venn diagram, then answer these questions. Which is the smallest group of patterned fish? Which chart shows this information most clearly?



Answers: 1) four 2) two 3) stripy and spotty fish, Venn diagram

## What colour are the fish?

The charts and graphs below all show how many red, green and yellow fish there are. Look at the picture to answer the questions, then check that each chart is correct.

How many red fish are there?

**Tally chart**

| | |
|---|---|
| red fish | III |
| green fish | II |
| yellow fish | IIII |

**2** How many yellow fish are there?

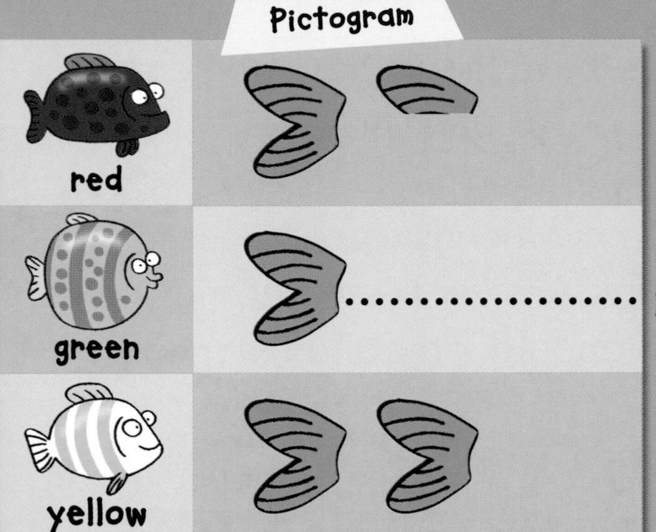


**3** Which is the biggest group of fish colours?

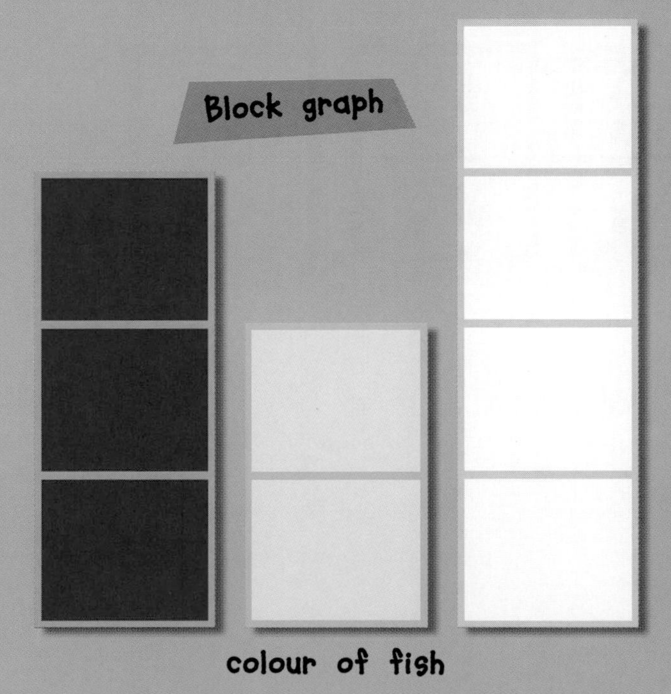


**4** Now look at these charts one by one and answer the following questions. How many more red fish are there than green fish? Which chart gives you the answer quickest?

Answers: 1) three 2) four 3) yellow 4) one, block graph

# Test your knowledge

Now you have learnt about different kinds of charts and graphs, it's time to test your chart-reading knowledge. Harry's new friends want to get to know him. Answer these questions to find out about Harry and his friends.

### 2. Harry's home

Harry's new friends all live on his street. Their house numbers are shown on this list. Harry lives at number 39. Who is his closest neighbour?

### 1. School star

The children's stars for last term are shown on this tally chart. Who shares Harry's score?

| | |
|---|---|
| Harry | \|\|\|\| |
| Carla | ~~\|\|\|\|~~ \|\|\| |
| Abdul | \|\|\|\| |
| Sonny | \|\|\| |
| Mary | \|\| |
| Nick | ~~\|\|\|\|~~ ~~\|\|\|\|~~ |

### 3. Fave food!

On this block graph, you can see the children's favourite foods. Harry doesn't like icecream or hamburgers. What food does he like best?

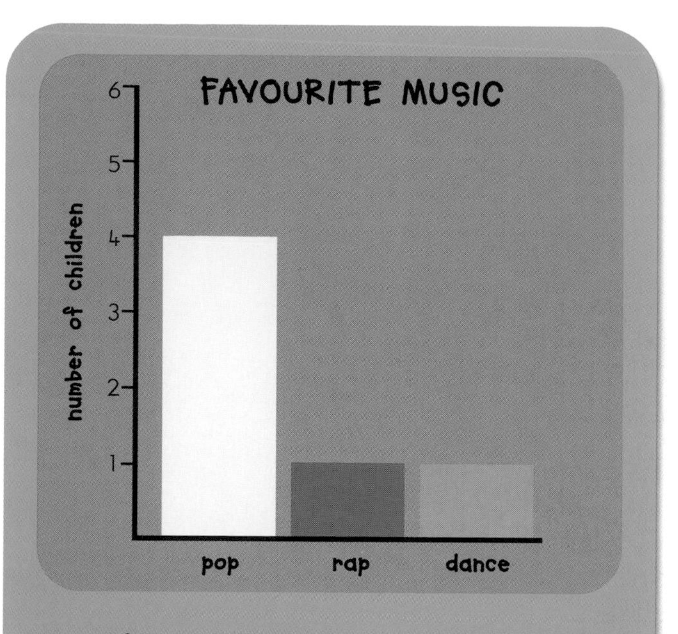


## 4. Music mad

This bar chart shows the children's favourite music. Harry likes the same music as most of his friends. What kind does he like best?

| | Has sister | Has no sister |
|---|---|---|
| Has no brother | Nick | Abdul, Sonny |
| Has brother | Carla | Harry, Mary |

## 5. Family matters

Here's a Carroll diagram showing which of the children have brothers and sisters. Does Harry have a sister?

# Top tip

Don't forget to label your graphs and give them a title. This is so that people who read them will know what information they show!

HARRY'S T-SHIRTS

number of t-shirts

4
3
2
1

red green blue

colour of t-shirts

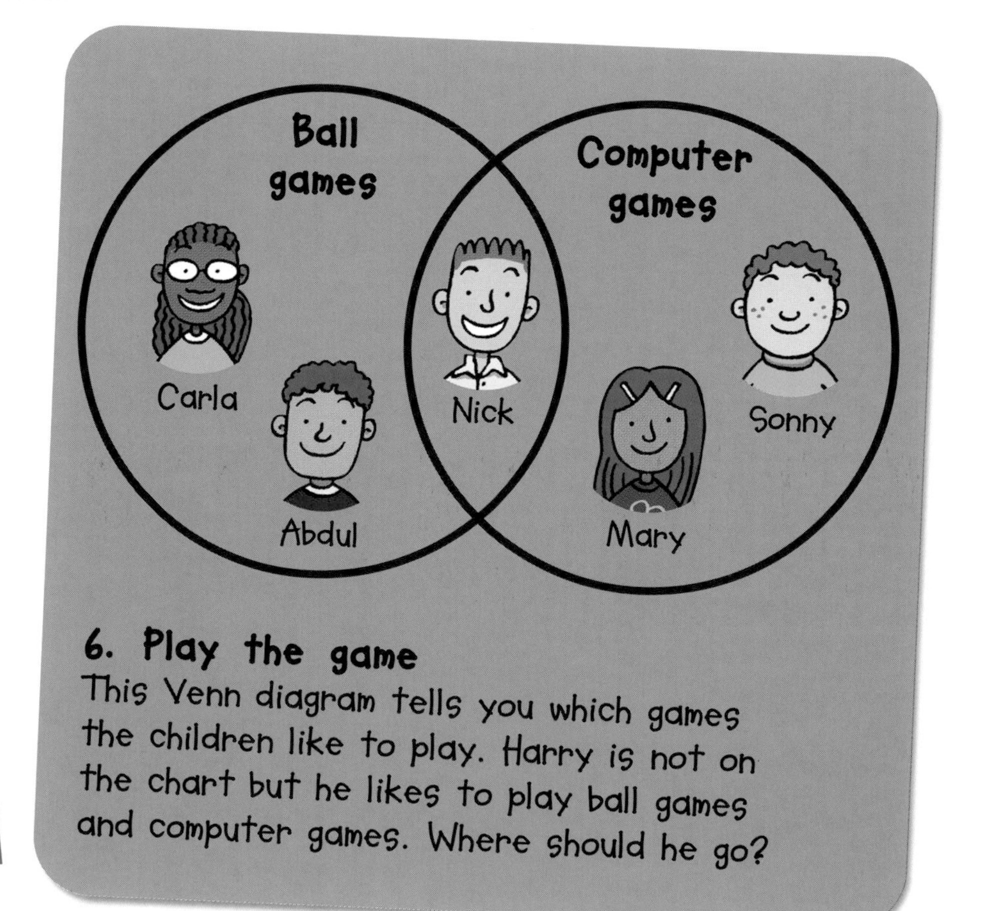


## 6. Play the game

This Venn diagram tells you which games the children like to play. Harry is not on the chart but he likes to play ball games and computer games. Where should he go?

Answers: 1) Abdul 2) Sonny 3) chips 4) pop 5) no 6) in the middle set with Nick

# Useful words

**axis**
This is a line on a graph which is labelled with information. There are always two axes, one along the bottom of the graph and one along the left-hand side.

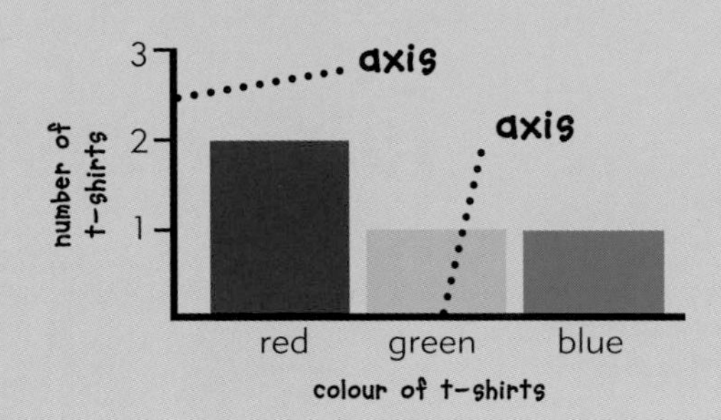


**bar chart**
This chart shows information in bars. To read a bar chart, look at the height of a bar against the number shown on the axis. You can draw bars up or across the page.

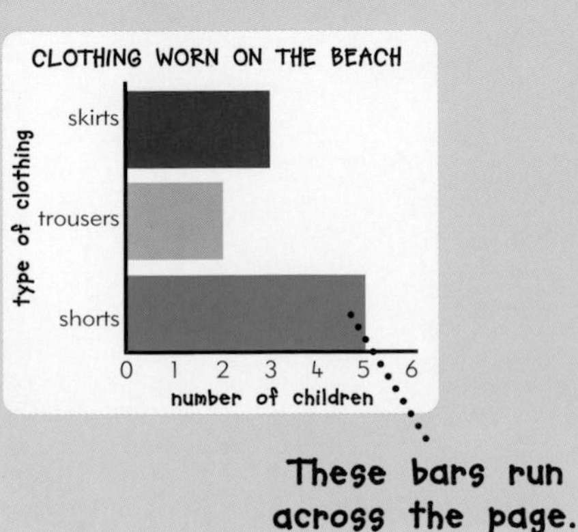


**block graph**
This chart shows information in separate blocks which are stacked into piles. To find out how many items there are in a pile, count the number of blocks.

**calendar**
A calendar is a chart which shows the date of each day in one particular year. There is one page for each month.

AUGUST

| Monday | Tuesday | Wednesday | Thursday | Friday | Saturday | Sunday |
|---|---|---|---|---|---|---|
| | | | 1 | 2 | 3 | 4 |
| 5 | 6 | 7 | 8 | 9 | 10 | 11 |
| 12 | 13 | 14 | 15 | 16 | 17 | 18 |
| 19 | 20 | 21 | 22 | 23 | 24 | 25 |
| 26 | 27 | 28 | 29 | 30 | 31 | |

Dates run across in rows.

**Carroll diagram**
A Carroll diagram compares different kinds of information at the same time.

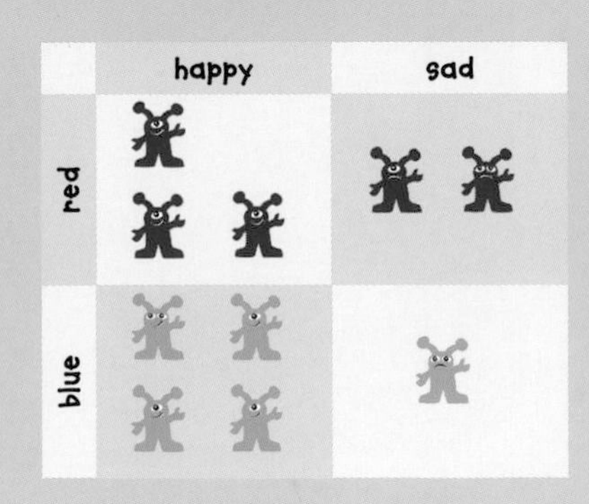


**chart**
A chart is a table or a picture, such as pictogram, that shows information in rows and columns.

**column**
A column is set out down the page.

**digit**
This is the symbol we use to stand for a number. **0, 1, 2, 3, 4, 5, 6, 7, 8, 9** are digits. The number **12** has two digits, **1** and **2**.

**frequency**
This means how many times a particular event occurs.

**frequency table**
A frequency table shows you how often a particular event happens over a period of time.

| | Frequency | Total |
|---|---|---|
| rainy days | \|\|\| | 3 |
| sunny days | ~~\|\|\|\|~~ \|\| | 7 |

**graph**
This is a diagram with blocks, bars or lines which compares two kinds of information. A block graph is one kind of graph.

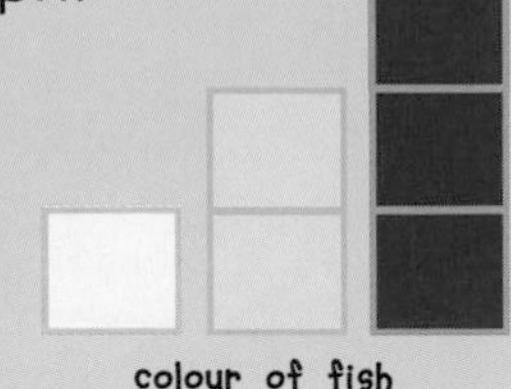

**list**
A list is a group of words or items written one below the other, such as a shopping list.

**pictogram**
This is a chart which shows information in pictures, called symbols, rather than words or numbers.

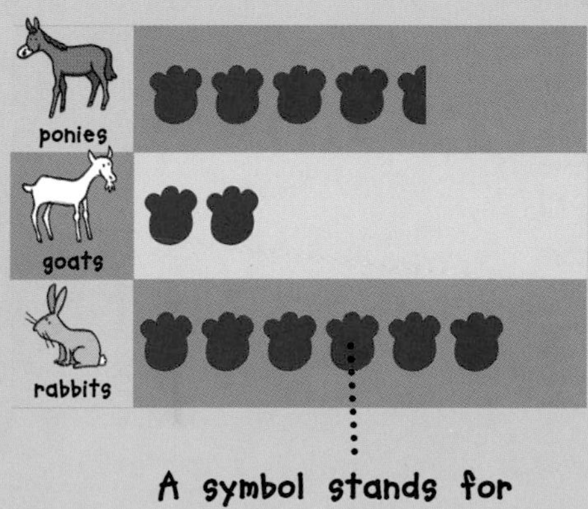


**row**
A row is set out across the page.

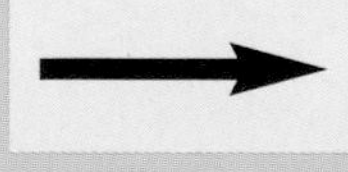

**set**
When items can be grouped together because they are similar, we say that they make a set.

**symbol**
In a pictogram, a picture called a symbol is used to stand for a number of objects.

**table**
This is a set of facts or numbers set out clearly in rows or columns.

SCORE TABLE

| | Tom | Susy |
|---|---|---|
| First roll | 30 | 30 |
| Second roll | 10 | 50 |
| Third roll | | |
| Total score | | |

**tally chart**
Keeping a tally is a way of counting. In a tally chart, marks are used instead of numbers to keep score.

<table>
<tr><td>Joe</td><td>|</td></tr>
<tr><td>Kate</td><td>||</td></tr>
<tr><td>Liz</td><td>卌</td></tr>
</table>

These marks stand for the number five.

**Venn diagram**
A Venn diagram groups objects that are similar into sets which you can then compare.

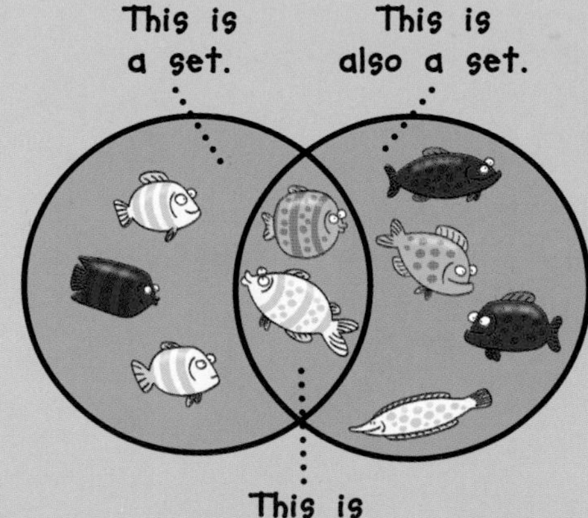


# Index

# Notes for parents and teachers

**This book helps children to understand information that is shown as a chart or a graph. Sometimes we create charts and graphs to help us organise our thoughts. More often, we need to interpret and understand the charts and graphs we see around us. Shopping lists, bus and train timetables, and calendars are all examples of charts that children might see every day.**

## Make a list

Show your children some of the ways you make lists to help you in everyday life.

- Ask children to make lists of their own. What would they buy from the supermarket? Can they list all their toys from memory?

- Can children put their lists into alphabetical order? Does this make the lists easier to understand?

## Count the days

Reading a calendar is an important life skill, but designing one is lots of fun!

- Draw month grids on to coloured card and decorate them with paint, glitter or even family photos. You can also find calendar layouts on computer software which you can use as templates.

## Picture it!

Pictograms are fun to learn about because they are visual.

- Ask children to create pictograms like the ones on pages 12 and 13. For example: How many hours of TV did children watch at the weekend? One TV symbol equals 2 hours!

## Block graphs

Block graphs are easy to understand because each block represents one item.

- Invite children to make a block graph of their friends' favourite cartoon characters. They can draw cartoons to stick on the squares.

## Bar charts

Bar charts enable us to compare totals quickly by comparing the height of each bar.

- Ask children to invent new challenges from the bar charts on pages 16 and 17. Here are a few examples. How many more kids are wearing shorts than trousers? Which beach is best for surfing?

## Venn diagrams

Venn diagrams help children to develop logical thinking by working out how things can be grouped together.

- Ask children to look at a collection of household items and find different ways of sorting them out. For example: What is the item used for? What is it made of? Then ask them which items fall into more than one group.

## Tally it up

Tally charts and frequency tables can be used to record the results of surveys and experiments.

- Create a tally chart to record the number of vehicles that pass your window in a 10-minute period. Write down how many buses, cars and bicycles pass by. Then tally up the totals and make a frequency table.

## Make it fun!

Charts and graphs provide plenty of opportunity to be colourful, creative and to enjoy learning. Encourage children by praising their efforts to learn – and always finish on a correct answer to keep their attitudes positive and their confidence levels high.